A PEEK AT STEMS

Published in 2026 by The Rosen Publishing Group, Inc.
2544 Clinton Street, Buffalo, NY 14224

First Edition

Editor: Greg Roza
Book Design: Leslie Taylor

Photo Credits: Cover, p. 1 Akintevs/Shutterstock.com; p. 5 brgfx/Shutterstock.com; p. 7 Elizabeth Foster/Shutterstock.com; p. 9 (ivy) Tadeas Skuhra/Shutterstock.com, (cucumber) Photo Win1/Shutterstock.com, (watermelon) Marina_vert/Shutterstock.com, (grapes) Mark Borbely/Shutterstock.com; p. 11 Dorling Kindersley ltd/Alamy.com, (inset) GraphicsRF.com/Shutterstock.com; p. 13 Dolfilms/Shutterstock.com; p. 15 Gerrit van den Bosch/Shutterstock.com; p. 17 (cactus) Brenda Brooks/Shutterstock.com, (potatoes) Dmitri Malyshev/Shutterstock.com, (moss) xpixel/Shutterstock.com, (water lily) Ruslan Suseynov/Shutterstock.com; p. 19 (Asparagus) Africa Studio/Shutterstock.com, (broccoli) barmalini/Shutterstock.com, (tree farm) Danielle MacInnes/Shutterstock.com, (wood) Viktor Sergeevich/Shutterstock.com; p. 21 DC Studio/Shutterstock.com.

Cataloging-in-Publication Data

Names: Bradley, Doug, 1971-.
Title: A peek at stems / Doug Bradley.
Description: Buffalo, New York : PowerKids Press, 2026. | Series: A peek at plants | Includes glossary and index.
Identifiers: ISBN 9781499452068 (pbk.) | ISBN 9781499452075 (library bound) | ISBN 9781499452082 (ebook)
Subjects: LCSH: Stems (Botany)–Juvenile literature.
Classification: LCC QK646.B696 2026 | DDC 581.4'95–dc23

Manufactured in the United States of America

CPSIA Compliance Information: Batch #CSPK26. For Further Information contact Rosen Publishing at 1-800-237-9932.

Find us on 

CONTENTS

What Do Stems Do?

Have you ever thought about what different plant parts do? Roots take in water and **nutrients** from the soil. Flowers and fruits help make new seeds. Leaves take in sunlight to make food for the plant. But what do stems do? Let's find out!

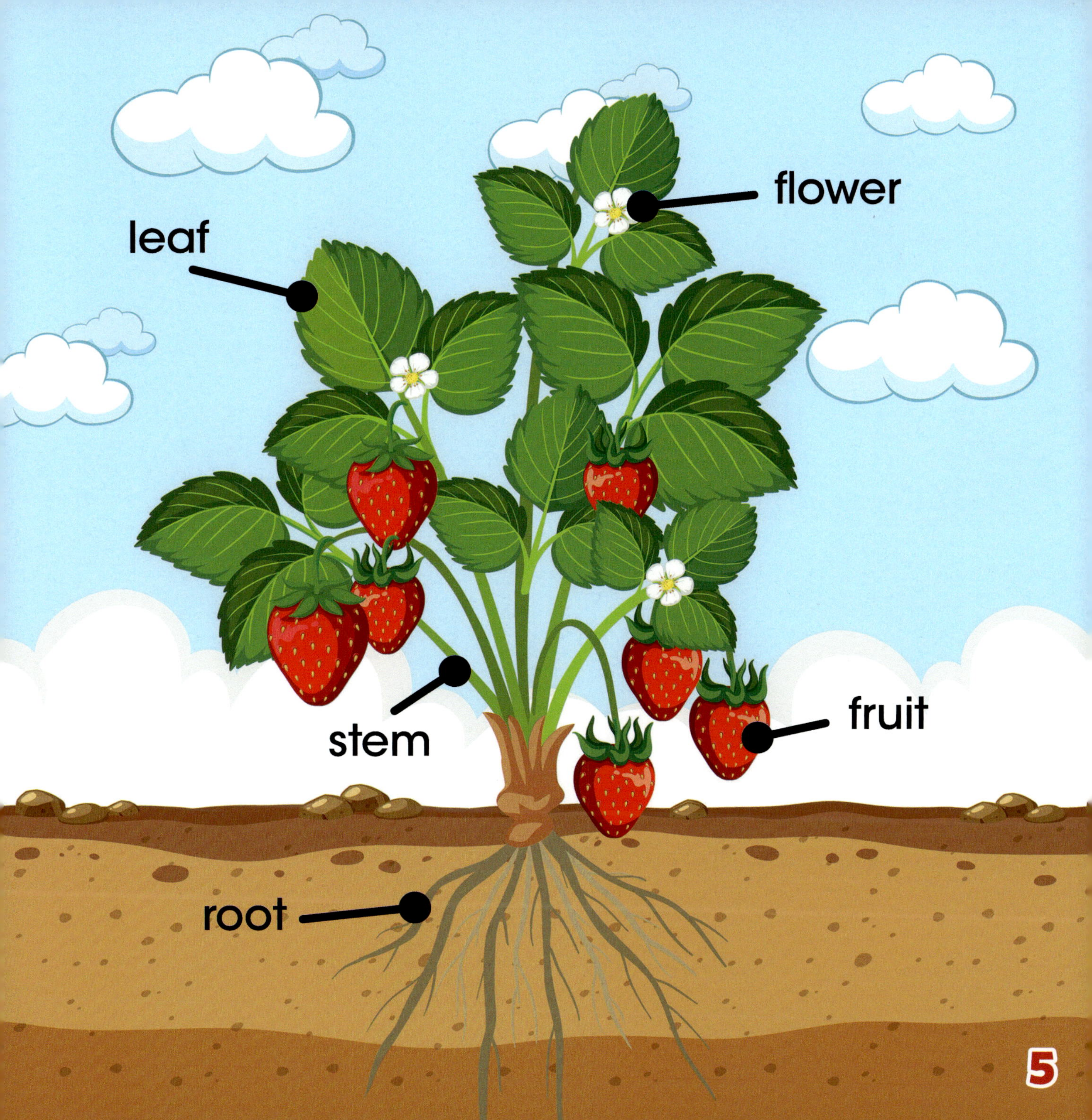

flower
leaf
stem
fruit
root
5

Branching Out

The stem is the main body of a plant. It holds the plant up. Stems grow taller. Some grow really tall! Plant stems often split into smaller stems called branches. Branches can split into even smaller branches. Leaves, fruit, and flowers grow on stems and branches.

7

Vines

Vines are stems. Vines are thin and bendable. Vines grow quickly and **curl** around other things. This lets vines "climb" up high. Ivy is a plant that climbs walls. Some food grows on vines. Cucumbers, watermelons, and grapes grow on vines.

ivy
cucumbers
watermelon
grapes

Inside the Stem

Plants need water and nutrients to grow. Roots take them from the soil. Then they move up the stem through tiny **tubes**. They pass from the stem into the leaves, flowers, and fruits. Water and nutrients are sometimes stored in the stem for later use.

INSIDE STEMS

Plant Skin

The outside of a stem keeps the inside safe. It also helps keep water inside the stem. Woody plants, or trees, have a hard outer **skin** called bark. Bark keeps trees safe from many things, such as bugs, birds, and weather. However, beavers have no problem biting through bark!

13

Giant Stem!

The largest stems in the world are in California. General Sherman is a giant sequoia tree that is 2,200 years old. It is 275 feet (83.8 m) tall. General Sherman's **trunk** is about 102 feet (31 m) around. That makes it the largest stem in the world!

GENERAL SHERMAN
15

Strange Stems

Most stems are long and have branches. Some stems are different. Barrel cactuses have ball-shaped stems. Potatoes are stems that grow under the ground. Water lilies have stems that grow under the water *and* under the ground.

barrel cactus

potatoes

water lily

Stems on the Farm

Farms grow many kinds of stems that we eat. Asparagus and broccoli are stem vegetables. Some farms grow trees for their wood. People use wood to build many things. Farms that grow fruit trees are called orchards. Some farms grow and sell Christmas trees!

asparagus

broccoli

tree farm

wood

Stems in the Garden

All gardens have stems, and some stems are good to eat! You can make a garden with your favorite stem vegetables. Asparagus and broccoli are good ones to choose. You can even plant your own tree! What kinds of stems would you plant in your garden?

GLOSSARY

curl: To form into or grow in circles or ringlets

nutrient: Something taken in by a plant or animal that helps it grow and stay healthy.

skin: The outer layer of a living thing.

trunk: The stem of a tree.

tube: A long, thin channel within a plant or animal that moves water and nutrients.

FOR MORE INFORMATION

BOOKS

Chu, Katherine. *Parts of a Tree.* Minneapois, MN: Learner Publishing, 2025.

Kirkman, Marissa. *Stems.* North Mankato, MN: Pebble, 2020.

Mihaly, Christy. *California's Redwood Forest.* Lake Elmo, MN: Focus Readers, 2018.

WEBSITES

Go Plant a Tree
www.youtube.com/watch?v=upUDTbBv1tA
This informative video shows a team of kids planting two trees while discussing tree science.

INDEX

TITLES IN THIS SERIES

A PEEK AT FLOWERS

A PEEK AT FRUITS

A PEEK AT LEAVES

A PEEK AT ROOTS

A PEEK AT STEMS

PowerKiDS
press™